*Timing is approximate. It may vary from place to place and year to year.

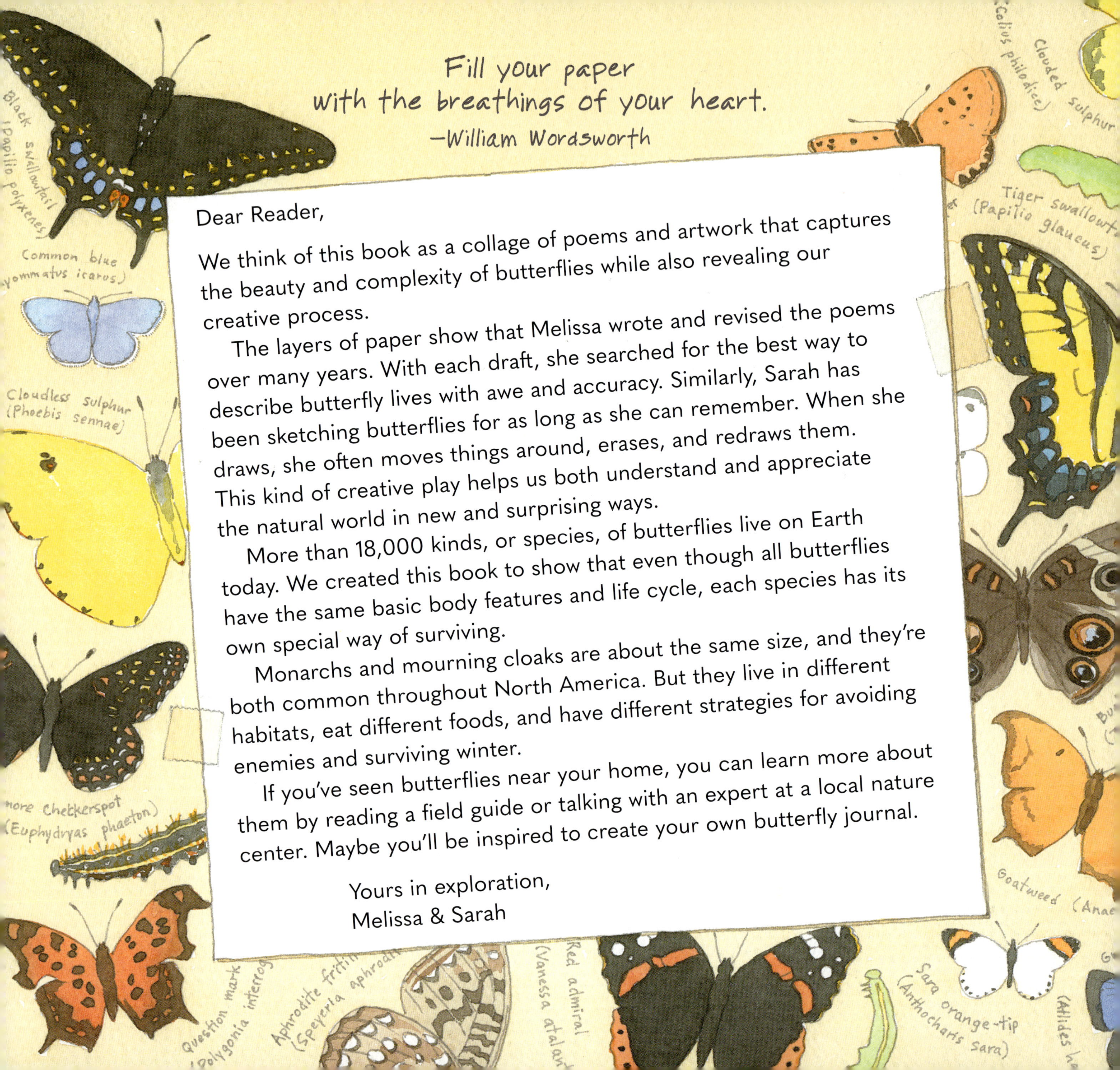

Fill your paper
with the breathings of your heart.
—William Wordsworth

Dear Reader,

We think of this book as a collage of poems and artwork that captures the beauty and complexity of butterflies while also revealing our creative process.

The layers of paper show that Melissa wrote and revised the poems over many years. With each draft, she searched for the best way to describe butterfly lives with awe and accuracy. Similarly, Sarah has been sketching butterflies for as long as she can remember. When she draws, she often moves things around, erases, and redraws them. This kind of creative play helps us both understand and appreciate the natural world in new and surprising ways.

More than 18,000 kinds, or species, of butterflies live on Earth today. We created this book to show that even though all butterflies have the same basic body features and life cycle, each species has its own special way of surviving.

Monarchs and mourning cloaks are about the same size, and they're both common throughout North America. But they live in different habitats, eat different foods, and have different strategies for avoiding enemies and surviving winter.

If you've seen butterflies near your home, you can learn more about them by reading a field guide or talking with an expert at a local nature center. Maybe you'll be inspired to create your own butterfly journal.

Yours in exploration,
Melissa & Sarah

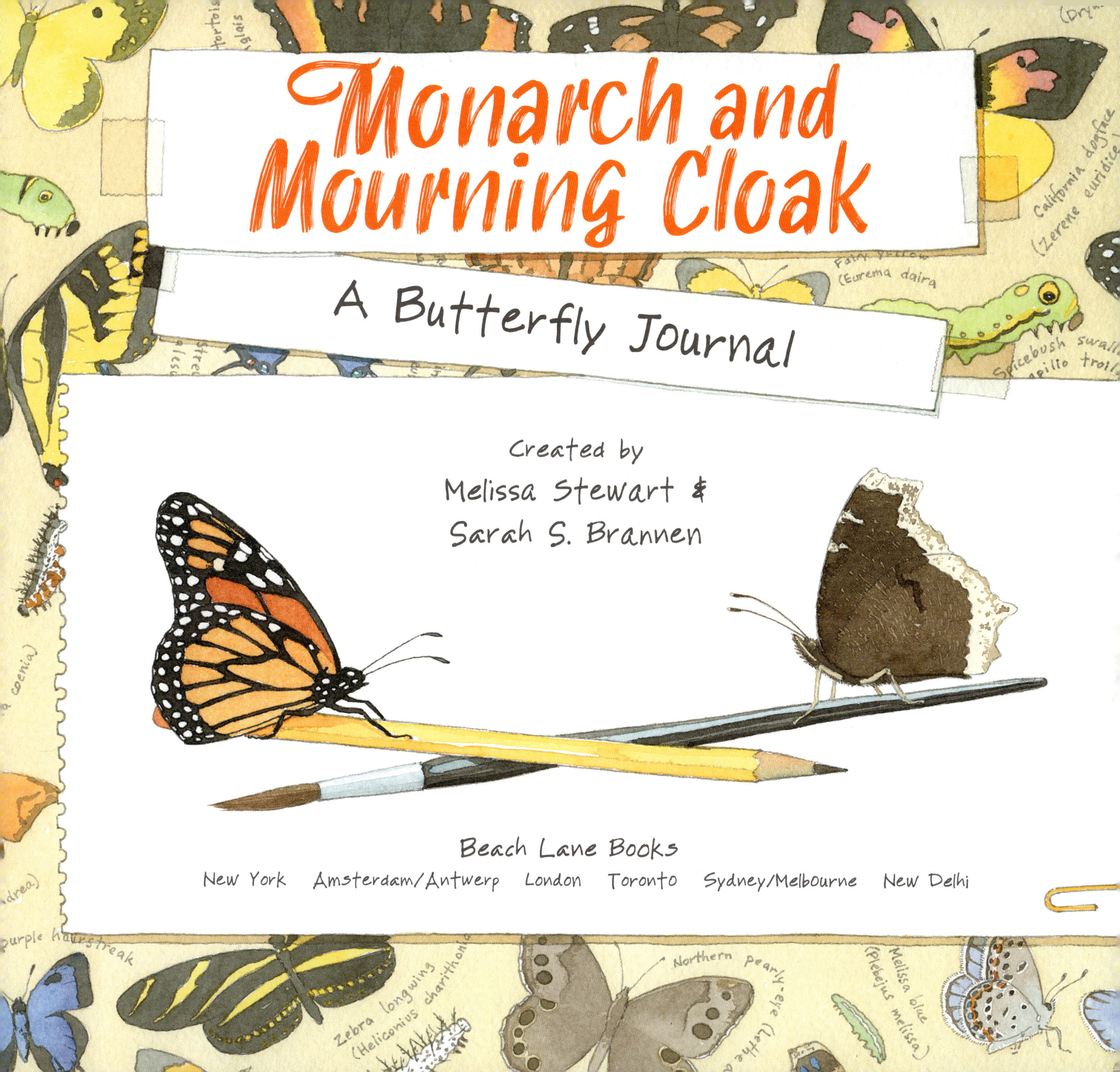
Monarch and Mourning Cloak
A Butterfly Journal
Created by
Melissa Stewart &
Sarah S. Brannen
Beach Lane Books
New York Amsterdam/Antwerp London Toronto Sydney/Melbourne New Delhi

Spotting Butterflies

Look, a monarch!

Bright orange wings
keep it safe.

They scream, "Poison!"

They stand out—
warning predators
to stay away.

There—a mourning cloak!

Dark brown wings
keep it safe.

They whisper, "Hidden."

They blend in—
tricking predators
so they stay away.

Butterfly Days: Monarch

Roaming,
rambling
above grass and flowers.

Searching,
seeking
fragrant blossoms.

Gliding,
sinking
toward its target—
sweet coneflower nectar.

June 19, 1:30 p.m.
Davis Field
A butterfly's antennae smell, touch, and hear.
daisy fleabane (Erigeron annuus)
yellow wild indigo
(Baptisia tinctoria)
honeybee
(Apis mellifera)

Yellow-bellied sapsucker
(Sphyrapicus varius)
Butterflies taste with their feet!

black-capped chickadee
(Poecile atricapillus)
Butterfly Days: Mourning Cloak
Dodging,
darting
between towering trees.
Searching,
seeking
tender stems.
Swerving,
curving
toward its target—
sugary oak sap.
June 21, 8:00 a.m.
Harvard Forest
white oak leaf

As Autumn Arrives

Monarchs gorge on goldenrod nectar,
storing fat
for the long, hard journey
to their warm southern home.

A mourning cloak sips sassafras sap,
storing fat
for a long, peaceful nap
in a safe, cozy spot.

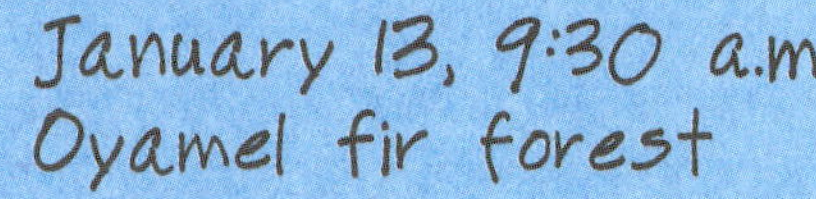

A Winter's Rest: Monarch

Packed tight,
in plain sight,
monarchs hibernate.

A Winter's Rest: Mourning Cloak

Tucked tight,
out of sight,
a mourning cloak hibernates.

Shhhh.

February 10, 3:30 p.m.
Wachusett Meadow
Wildlife Sanctuary

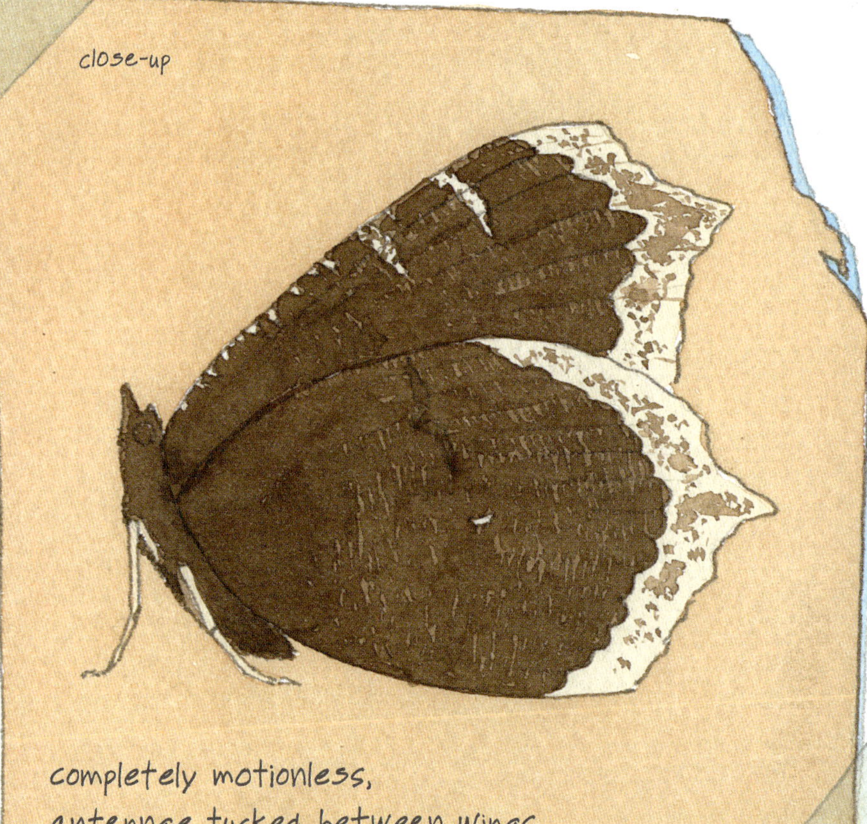

winterberry
(Ilex verticillata)
red fox (Vulpes vulpes)
hibernating
butterfly

The Promise of New Life

As spring returns,
monarchs meet
and mate.

The female lays eggs
on her northward journey.

She places them carefully,
one at a time,
under milkweed leaves.

Then she flies away.

As spring returns,
mourning cloaks meet
and mate.

The female lays eggs
in her year-round home.

She drops them hastily,
in large clumps,
around willow twigs.

Then she flies away.

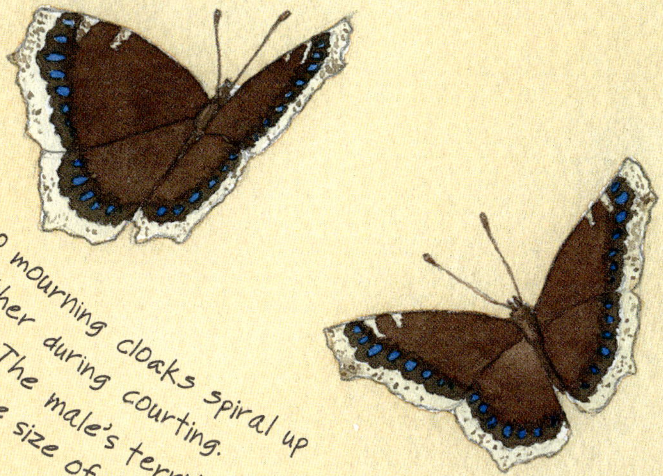
Two mourning cloaks spiral up
together during courting.
The male's territory is about
the size of a tennis court.

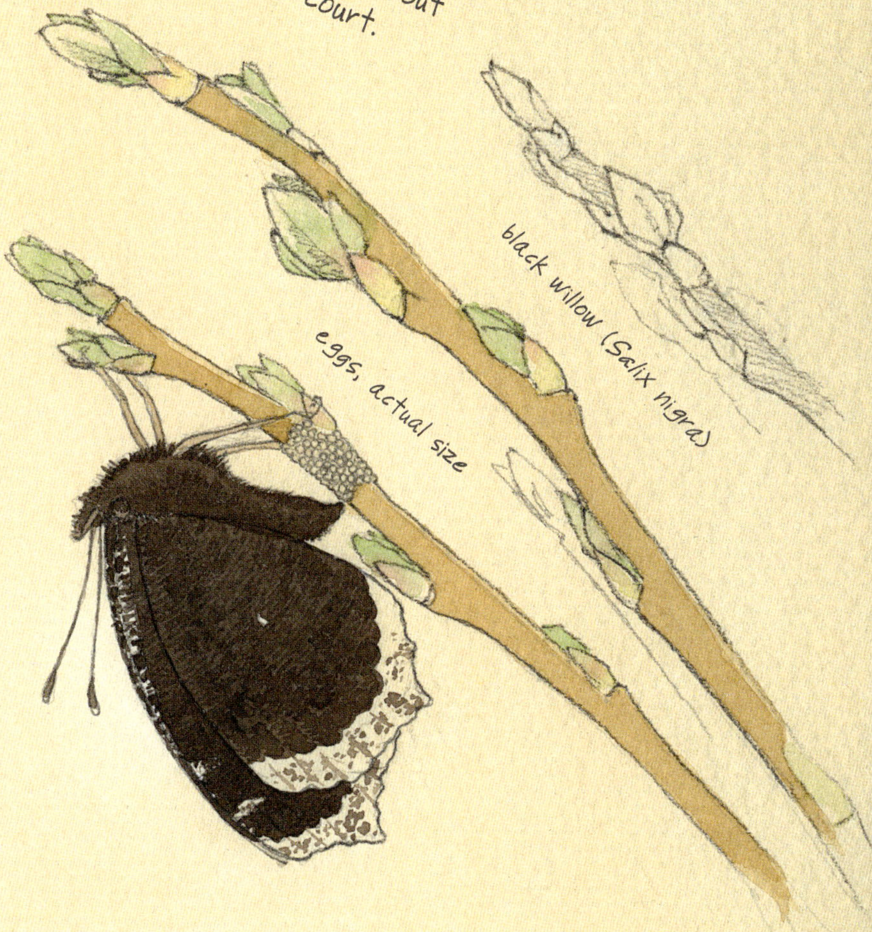
black willow (Salix nigra)
eggs, actual size

egg mass
enlarged

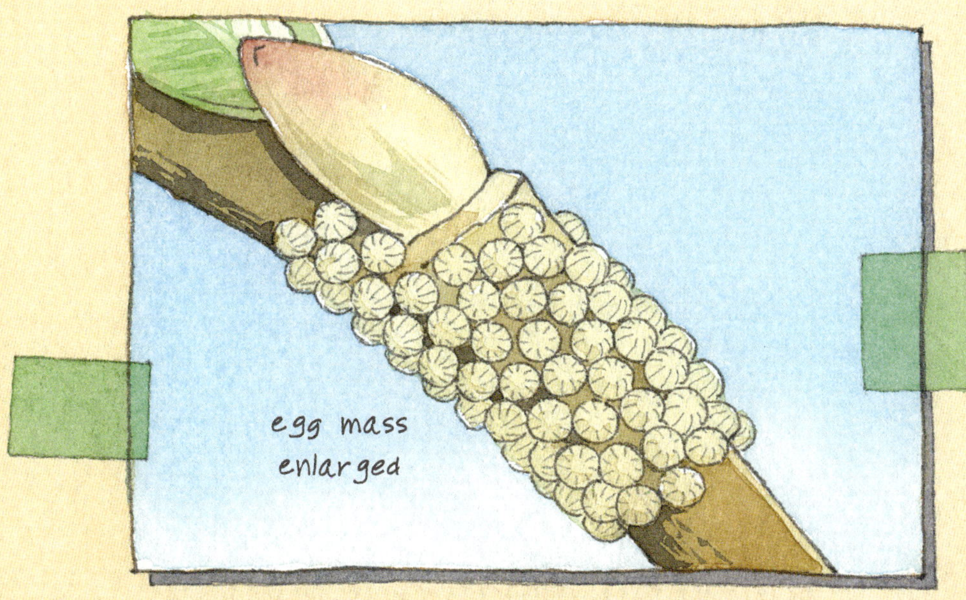

March 19, 11:15 a.m.
Broad Meadow Brook Wildlife Sanctuary

Egg-mergence

Sheltered.
Wrapped.
Wedged.
Trapped.

Tiny larvae are ready,
so ready,
to enter the world.

They wriggle and nibble.
Then chomp, chomp, chew.

They push and squirm.
Then pull, pull, pull—
until finally . . .

they're free!

The mourning cloak lays her eggs near new leaves so the caterpillars will have food.

Monarch

Day 1-4
iridescent shell

Day 5
black head is visible

Mourning Cloak

actual size

Hatching takes less than 3 minutes.

Day 5, continued . . .
larva nibbles hole in shell

crawls out of shell

eats shell

to reddish to brown, almost black.

Eggs hatch after 5-9 days.

actual size

Caterpillar Days

A monarch eats alone,
all by itself,
a peaceful milkweed meal.

A monarch faces danger alone,
all by itself,
as its colors shout: "Alert!"

July 19, 9:45 a.m.
Sarah's garden

Some hunters miss
these warning signs,
and, oh, do they pay the price.

Just one bite of

bitter flesh—

American robin (*Turdus migratorius*)

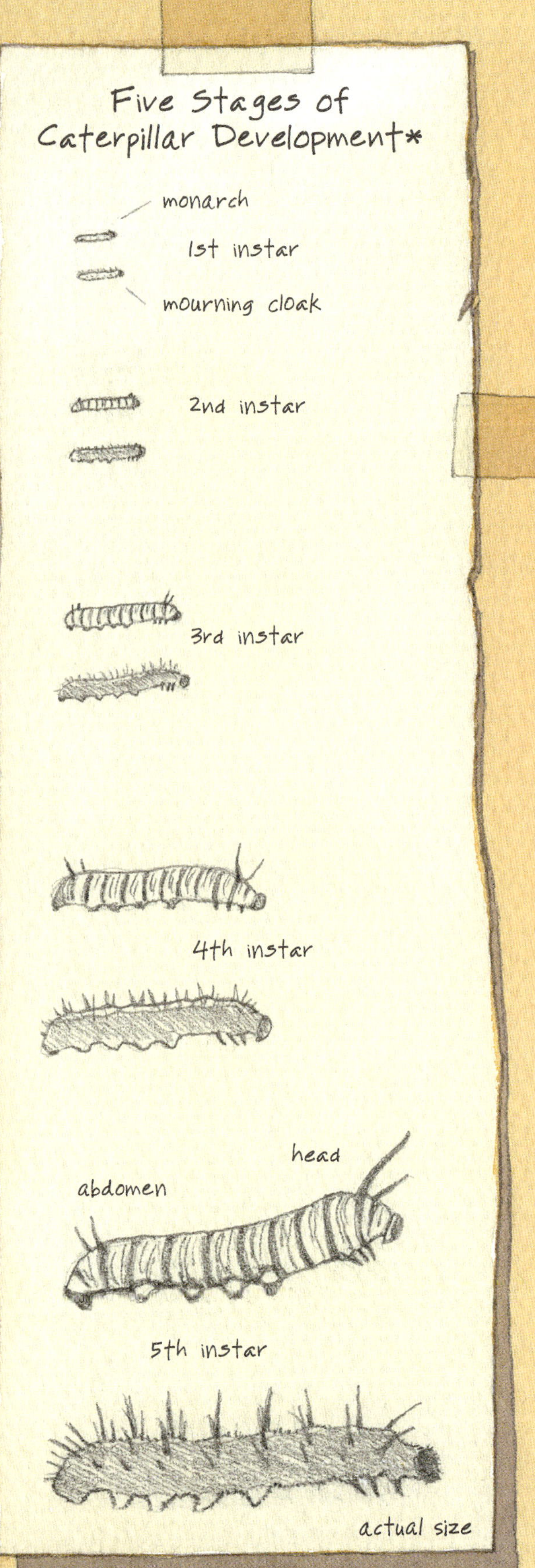

Munching and Molting

Day after day,
rain or shine,
caterpillars keep on
munching-crunching,
greedily gobbling.

They grow
and grow
and grow

until
bulging bodies
press against
stiff skins.

Hour after hour,
caterpillars stay
stock-still.

*Growth stages between molts are called instars.

July 23, 8:00 a.m.
Inside Sarah's studio

Then,
muscles contract
with rippling rhythms!

As caterpillars thrust their thoraxes
and drag their abdomens,
soft, moist underskins
slowly slide out of
worn, torn empty shells.

Molting starts at the head and moves backward along the body.

Then, to free its back end from the old skin, the caterpillar lifts its abdomen and waves it around.

Caterpillar Clocks

After a few weeks
   of gorging and growing,
   munching and molting,
caterpillars suddenly sense . . .

IT'S TIME!

It's time to find
   a safe, sturdy stem
high off the ground.

Monarch

The caterpillar hangs for 12–18 hours. Then rippling contractions start and last for an hour or so.

The pupa thrashes to get rid of the shed skin.

It's time to spin
a strong, silky pad
and hang head down.

It's time to shed
one last larval skin.
It's time to become a pupa.

Mourning cloak

Monarch:
Pupating takes 5-15 days, depending on temperature.

Mourning Cloak:
Pupating takes 10-15 days.

You can clearly see the wings and body.
Ready to emerge!

Changing, Rearranging

On the outside,
chrysalises seem silent and still.

But inside,
everything's changing, rearranging.

Heads shrink,
eyesight improves.

Antennae appear,
and coiled sipping straws
replace jaws.

Legs stretch l-o-n-g-e-r,
and pupae
grow
WINGS.

Can you see them?

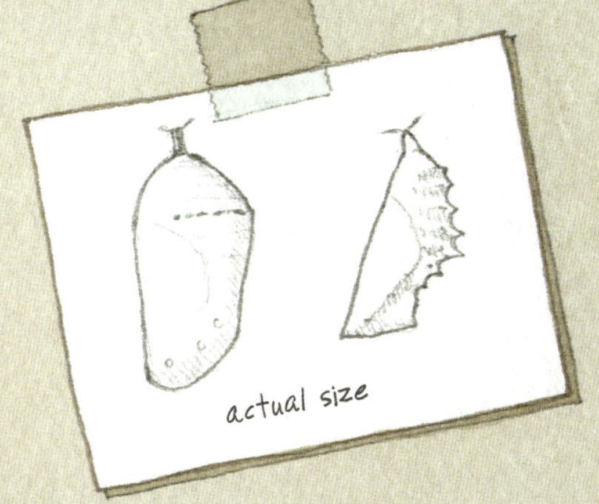

August 7, 2:00 p.m.
Outside Sarah's studio

Pushes and rests for about 90 seconds. Then the abdomen suddenly pops out . . .

. . . and swings down.

Break Out

Cramped.
Crumpled.
Wrinkled.
Rumpled.

Butterflies are ready,
so ready,
to make their escape.

As the summer sun rises
and a gentle breeze blows,
chrysalises crack.

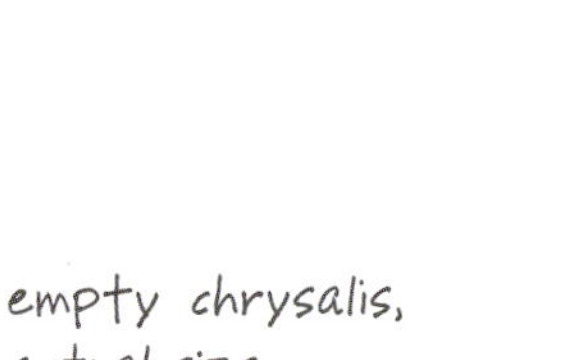

empty chrysalis, actual size

Wings are stretched out and flat in 10-20 minutes.

After 1-2 hours, starts opening and closing wings.

Butterflies push
and rest,
push
and rest,
as their bodies slide
down,
down,
and out
into the world.

Then fluid flows out
of their abdomens,
stretching wrinkled wings.

Slowly,
slowly,
their bodies harden
and wings spread wider,
longer,
flatter.

Until finally . . .

September 7, 1:30 p.m.
Along the edge of
Lindentree Farm

butterflies soar into the sky.

The Joy of Journaling

There's no right or wrong way to keep a journal. What you focus on, how you organize it, and how you record information is up to you.

I've been keeping nature journals since I was 20 years old, inspired by a friend who showed me his journal. I use notebooks with lined pages because I write more than I draw. Sometimes I keep a leaf or a flower and press it inside the pages. I often draw maps of areas where I'm hiking and mark the spots where I see something special.

Keeping a record of my experiences and observations while I'm on outdoor adventures helps me pay close attention to my surroundings and immerse myself in the wonder of the natural world.

—Melissa Stewart

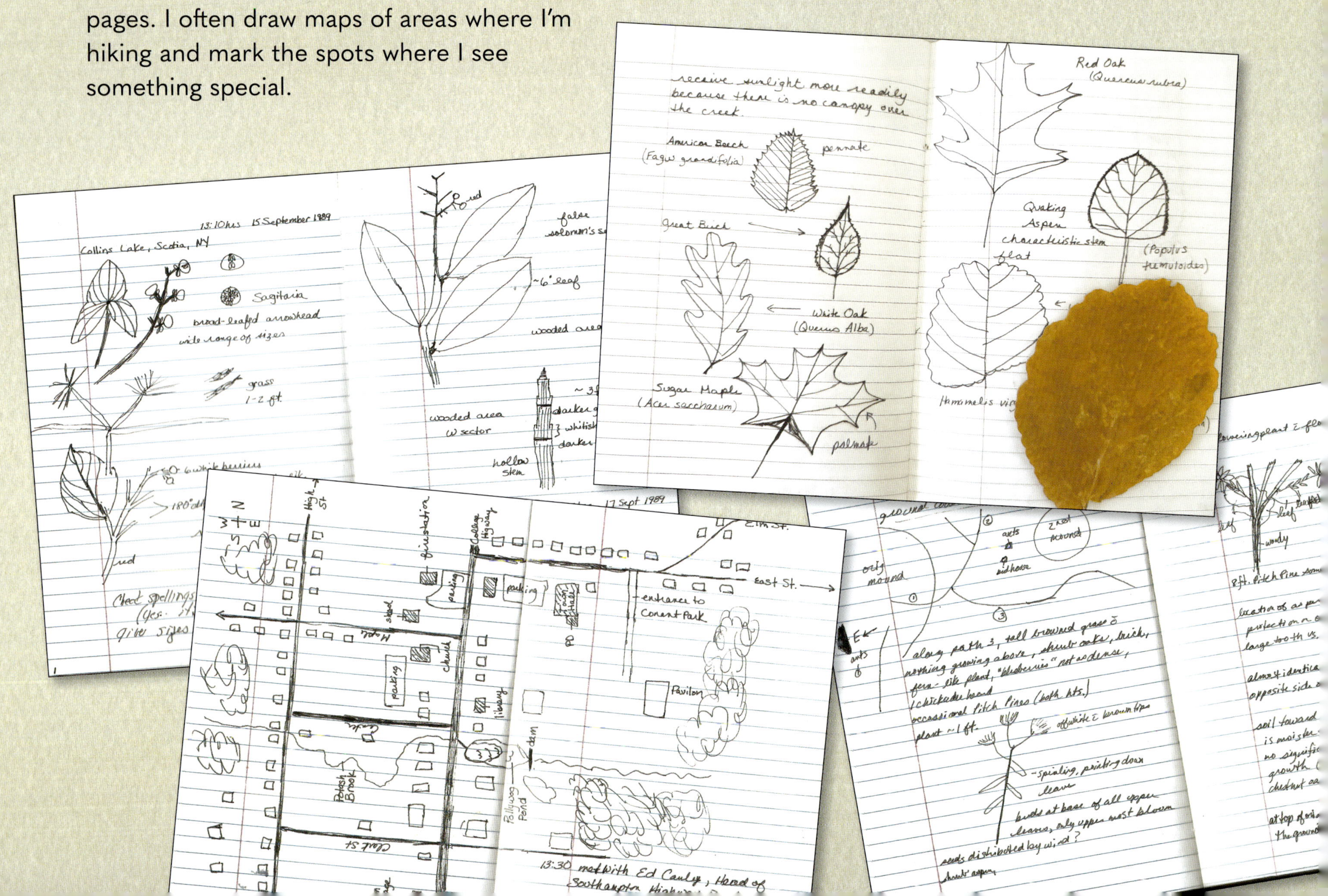

Creating a Sketchbook

My parents gave me a field guide to butterflies when I was 6 years old, and it quickly became my favorite book. I read it so many times that it fell apart. A few years later, a teacher gave me my first sketchbook. Even the look of the book, with its plain cover and blank white pages, was exciting. I carried it with me everywhere.

If you'd like to try keeping a sketchbook, all you need is a book with blank pages and something to draw with. You don't need fancy paints or brushes. Then just go outside and *look*.

When you sketch something, you look harder at it than you ever imagined you could. Don't worry about erasing—mistakes happen. Note the place and time and whatever you know about the thing you're drawing.

I often tape things into my sketchbook: a short article, a museum ticket, maybe a flower. And that's all there is to it. Whatever happens, have fun sketching!

—Sarah S. Brannen

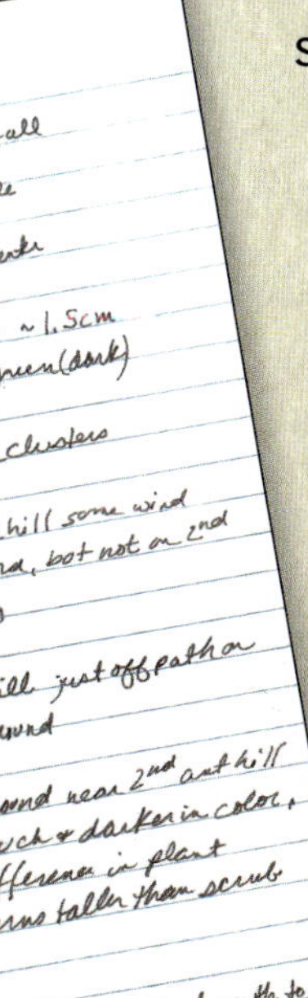

Butterfly Notes

Spotting Butterflies

Adult monarchs and mourning cloaks have different ways of staying safe.

When monarchs are in the caterpillar stage, they eat milkweed. Toxins from the plant stay in a butterfly's body for its whole life. The toxins taste bad to some animals and are poisonous to others.

mourning cloak playing dead

Mourning cloaks aren't poisonous. But their dark wings make them hard to spot when they perch on tree trunks or fallen logs. If a fast-flying bird gets too close, a mourning cloak can play dead. It drops to the forest floor and blends in with the leaf litter.

Butterfly Days

Monarchs and mourning cloaks live in different places. And they eat different foods.

Monarchs glide gracefully above fields, gardens, and meadows. They sip nectar from the large, colorful flowers growing in these wide-open habitats.

Mourning cloaks deftly dart through woodlands and wetlands. Large flowers full of nectar are rare in these habitats, so mourning cloaks usually feed on tree sap and rotting fruit.

mourning cloak feeding on rotting fox grapes (*Vitis labrusca*)

As Autumn Arrives

Monarchs and mourning cloaks prepare for chilly days in different ways.

Monarchs guzzle nectar before they migrate. Then they feed daily along the way. In North America, most monarchs east of the Rocky Mountains fly to mountain forests in Central Mexico. Some travel up to 3,000 miles (4,800 km). In the West, most monarchs fly to the California coast.

After their autumn feast, mourning cloaks search for shelter. Some slip into a rocky crevice. Others nestle under bark.

Monarch Butterfly Migration in North America

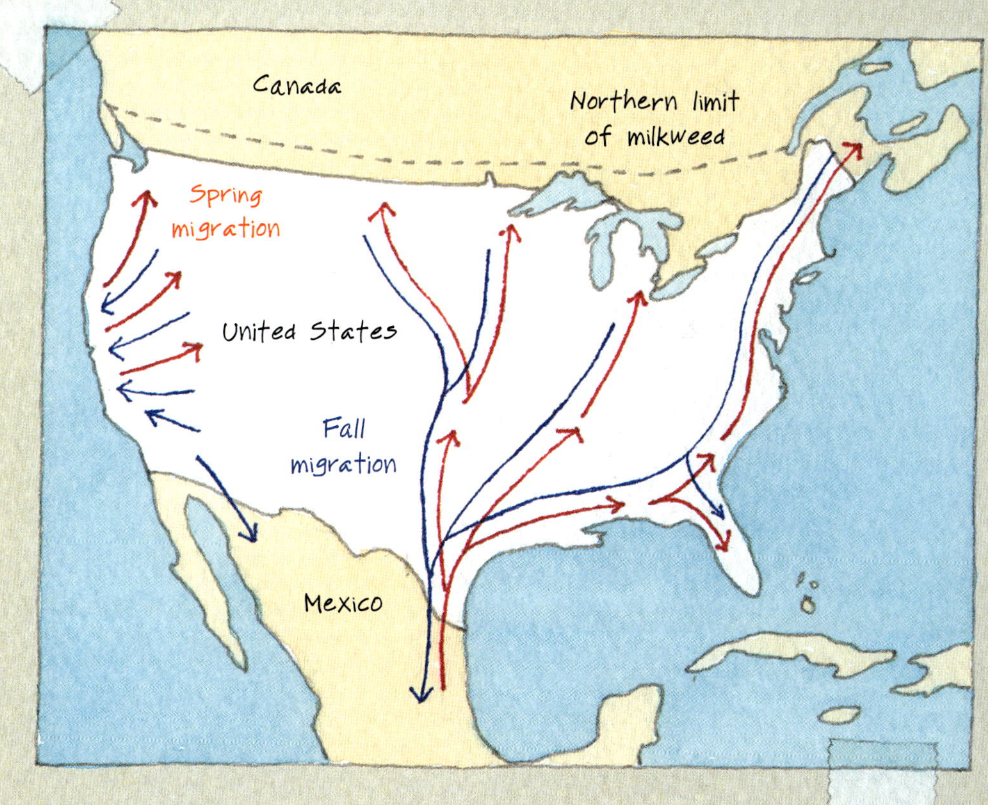

A Winter's Rest

Monarchs and mourning cloaks have different ways of surviving in winter.

On cool days, millions of monarchs cluster on trees and rest. Their hearts beat slowly, and they barely breathe. But on warmer days, the butterflies soak up the sun's heat until their muscles are ready to fly. Then the thirsty insects cruise through the sky in search of puddles and streams.

Week after week, month after month, a mourning cloak rests, silent and still. While it waits for spring, chemicals in its blood prevent its body from freezing. The butterfly can survive in temperatures as low as -22 °F (-30 °C).

monarchs sipping water

The Promise of New Life

Mating is similar in all butterflies, but monarchs and mourning cloaks lay eggs differently.

A mama monarch always lays eggs on milkweed. It's the only plant her young will eat. She lays 100 to 300 eggs, each on its own leaf. That way, the very hungry caterpillars won't have to compete for food. A mama mourning cloak lays up to 200 eggs in one place. She chooses willow, elm, birch, or cottonwood. These trees have lots of leaves, so there's plenty of food for the caterpillars.

monarch clinging to a milkweed leaf as it lays an egg on the underside

Egg-mergence

All butterflies begin their life inside an egg. Females lay lots of eggs, but many are eaten by predators, such as ladybugs, grasshoppers, earwigs, and ants. Caterpillars, pupae, and adult butterflies also face many dangers. Just 2 percent of all the eggs laid survive to adulthood.

A monarch egg hatches after 2 to 5 days. The first thing the caterpillar does is eat its own eggshell. It's full of protein and gives the little larva a good start in life.

A mourning cloak egg hatches after 5 to 9 days. The newly hatched caterpillar may eat the unhatched eggs of its brothers and sisters.

differential grasshopper (*Melanoplus differentialis*)

Caterpillar Days

Monarchs and mourning cloak caterpillars have different ways of staying safe.

The toxins in milkweed plants make monarch caterpillars taste bad. And they're poisonous to some animals. After even a tiny nibble, predators remember the caterpillar's striped body and never eat it again.

For mourning cloak caterpillars, there's safety in numbers. When they huddle together and twitch at the same time, many predators think they're one big, scary animal. If a hungry hunter does decide to attack, it gets a mouthful of spines. *Ouch!*

A Caterpillar's Body

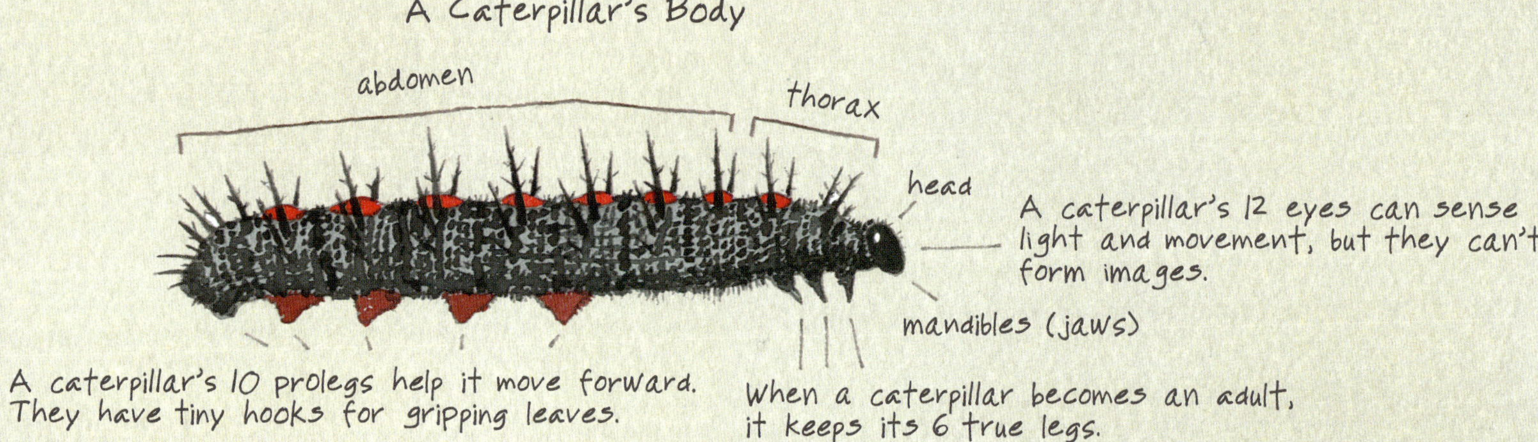

Munching and Molting

When a caterpillar outgrows its tough outer cuticle, it must molt, or shed its skin.

After breaking free of its old cuticle, a caterpillar rests for several hours. During this time, it gulps air to expand the new skin before it dries out and hardens. This gives the caterpillar room to grow. Finally, it eats the shed skin and pushes its old head capsule off the leaf.

A monarch spends about 2 weeks as a caterpillar. During that time, it grows 2,000 times bigger and 3,000 times heavier. It molts 5 times.

A mourning cloak spends about a month as a caterpillar. It also molts 5 times.

newly molted monarch caterpillar about to eat its old skin (cuticle)

Caterpillar Clocks

All caterpillars transform into pupae. Monarchs and mourning cloaks go through the process in the same way.

When they're fully grown, the caterpillars crawl to the ground. Then they wander in search of a sheltered spot where predators won't notice them. A monarch usually travels a distance about the length of a school bus. But a mourning cloak may wander four times as far.

After anchoring itself to a stem or a branch with strong silk threads from a gland below its mouthparts, a caterpillar hangs upside down. Several hours later, it molts one last time. As the insect wriggles and squirms, the old skin inches upward, revealing a chrysalis. The pupa attaches its stemlike cremaster to the silk pad. Then its old skin falls off. Finally, the pupa settles down, and its chrysalis skin begins to harden.

mourning cloak chrysalis hanging from a fox grapevine

Changing, Rearranging

All butterfly pupae go through incredible changes. Beneath the chrysalis skin, chemicals called enzymes dissolve some body parts into a soupy mixture. Then instructions in the insect's DNA guide the rebuilding process.

- The head shrinks and antennae grow.
- The twelve simple eyes are replaced by two compound eyes.
- The lips and jaws transform into a long, hollow tube.
- The intestines become shorter and simpler.
- The prolegs disappear, and the true legs grow longer.
- Buds of tissue in the thorax grow into wings.

The diagram below shows where some of the new body parts are inside the chrysalis.

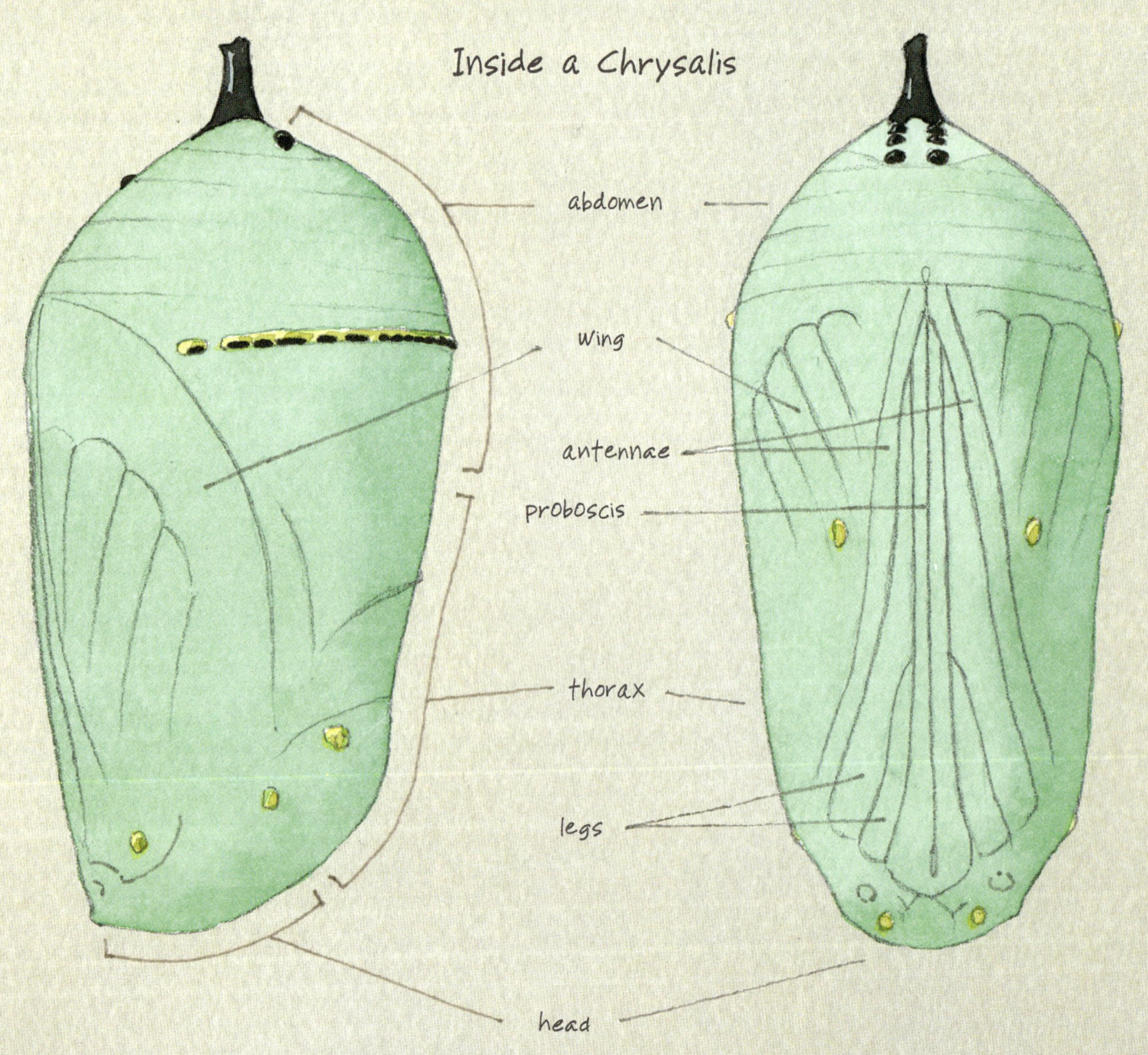

Break Out

Every butterfly breaks out of its chrysalis in the same way.

First, the butterfly's body separates from its chrysalis skin. Then the chrysalis cracks at the bottom, near the insect's head.

The butterfly pushes against the hard case to widen the crack. At the same time, it works to free its body, slipping down, bit by bit. Eventually, its swollen abdomen pops out and swings down, followed by tiny wings.

The butterfly grabs the outside of the case with its long legs, hauls itself up, and holds on tight. For the next few hours, it rests. As the butterfly's body dries and hardens, it pumps fluid out of its abdomen and into its wings. Its bulging body shrinks, and its wings expand.

Finally, it's ready to fly!

For Oona and Hattie.
May you grow up loving the natural world.
—M. S.

To my sister Jennie.
Thanks for being my flying buttress, boat mover,
and caterpillar wrangler.
—S. S. B.

We would like to thank Linda Graetz, former naturalist with Mass Audubon, and Tia Pinney, Mass Audubon senior naturalist and adult program coordinator, MetroWest, for their assistance with the illustrations.

BEACH LANE BOOKS
An imprint of Simon & Schuster Children's Publishing Division • 1230 Avenue of the Americas, New York, New York 10020 • For more than 100 years, Simon & Schuster has championed authors and the stories they create. By respecting the copyright of an author's intellectual property, you enable Simon & Schuster and the author to continue publishing exceptional books for years to come. We thank you for supporting the author's copyright by purchasing an authorized edition of this book. • No amount of this book may be reproduced or stored in any format, nor may it be uploaded to any website, database, language-learning model, or other repository, retrieval, or artificial intelligence system without express permission. All rights reserved. Inquiries may be directed to Simon & Schuster, 1230 Avenue of the Americas, New York, NY 10020 or permissions@simonandschuster.com. • Text © 2026 by Melissa Stewart • Illustration © 2026 by Sarah S. Brannen • Book design by Lauren Rille • All rights reserved, including the right of reproduction in whole or in part in any form. BEACH LANE BOOKS and colophon are trademarks of Simon & Schuster, LLC. • For information about special discounts for bulk purchases, please contact Simon & Schuster Special Sales at 1-866-506-1949 or business@simonandschuster.com. • Simon & Schuster strongly believes in freedom of expression and stands against censorship in all its forms. For more information, visit BooksBelong.com. The Simon & Schuster Speakers Bureau can bring authors to your live event. For more information or to book an event, contact the Simon & Schuster Speakers Bureau at 1-866-248-3049 or visit our website at www.simonspeakers.com. • The main text for this book was set in Quasimoda. The captions and headings were set in Butterfly Journal, a hand-lettered font created by the illustrator. • The illustrations for this book were rendered entirely by hand in watercolor, pen, and pencil on Arches cold press paper. • Manufactured in China
1025 SCP • First Edition
10 9 8 7 6 5 4 3 2 1
Library of Congress Cataloging-in-Publication Data • Names: Stewart, Melissa, author. | Brannen, Sarah S., illustrator. • Title: Monarch and mourning cloak : a butterfly journal / created by Melissa Stewart and Sarah S. Brannen. • Description: First edition. | New York : Beach Lane Books, [2026] | Audience: Ages 4–8 | Audience: Grades 2–3 | Summary: "Journal meets sketchbook in this luminously illustrated and thoroughly informative picture book poetry collection that captures the beauty and complexity of monarch and mourning cloak butterflies"—Provided by publisher. • Identifiers: LCCN 2025005403 (print) | LCCN 2025005404 (ebook) | ISBN 9781665962711 (hardcover) | ISBN 9781665962728 (ebook) • Subjects: LCSH: Monarch butterfly—Juvenile literature. | Mourning cloak butterfly—Juvenile literature. | Butterflies—Pictorial works. | Butterflies—Childrens' poetry. • Classification: LCC QL544.2 .S7453 2026 (print) | LCC QL544.2 (ebook) | DDC 595.78/9—dc23/eng /20250221 • LC record available at https://lccn.loc.gov/2025005403 • LC ebook record available at https://lccn.loc.gov/2025005404

Selected Sources

Baumle, Kylee. *The Monarch: Saving Our Most-Loved Butterfly*. Pittsburgh: St. Lynn's Press, 2017.

Bryant, Peter J. "Mourning Cloak Butterfly." Natural History of Orange County, California, and Nearby Places. School of Biological Sciences, University of California, Irvine. https://nathistoc.bio.uci.edu/lepidopt/nymph/mcloak.htm.

Burris, Judy, and Wayne Richards. *The Life Cycles of Butterflies: From Egg to Maturity, a Visual Guide to 23 Common Garden Butterflies*. North Adams, MA: Storey Publishing, 2006.

Collicutt, Doug. "Butterflies in the Snow," *NatureNorth*. http://www.naturenorth.com/spring/bug/mcloak/Mourning_Cloak.html.

Enroth, Chris. "The Mourning Cloak Butterfly," *Muddy River News*, February 25, 2023. https://muddyrivernews.com/markets-ag/the-mourning-cloak-butterfly/20230225083000/.

Hall, Donald W., and Jerry F. Butler. "Mourning Cloak Butterfly." *Featured Creatures*, University of Florida Institute of Food and Agricultural Sciences, February 2021. https://entnemdept.ufl.edu/creatures/bfly/mourning_cloak.htm.

The Incredible Journey of the Butterflies, directed by Nick de Pencier (Boston: PBS Video, 2009), DVD.

James, David G., ed. *The Book of Caterpillars: A Life-Size Guide to Six Hundred Species from Around the World*. Chicago: University of Chicago Press, 2017.

Leslie, Clare Walker. *Keeping a Nature Journal: Deepen Your Connection with the Natural World All Around You*. 3rd ed. North Adams, MA: Storey Publishing, 2021.

Nijhuis, Michelle. "Flight of the Monarchs." *National Geographic*, January 2024, pp. 36–67.

Sourakov, Andrei. "Monarch Butterfly." *Featured Creatures*, University of Florida Institute of Food and Agricultural Sciences, April 2021. https://entnemdept.ufl.edu/creatures/bfly/monarch.htm.

Stewart, Melissa. Personal observations recorded in nature journals, 1989–present.

University of Michigan's Animal Diversity Web. https://animaldiversity.org.

For Further Information

Aston, Dianna Hutts. *A Butterfly Is Patient*. San Francisco: Chronicle Books, 2011.

Delano, Marfé Ferguson. *Butterflies*. Washington, DC: National Geographic Kids, 2014.

Heiligman, Deborah. *From Caterpillar to Butterfly*. New York: HarperCollins, 2015.

North American Butterfly Association. http://naba.org.

Simon, Seymour. *Butterflies*. New York: HarperCollins, 2011.

Stewart, Melissa. *A Place for Butterflies*. 3rd ed. Atlanta, GA: Peachtree, 2024.

Wright, Amy Bartlett. *Peterson First Guide to Caterpillars of North America*. Boston: Houghton Mifflin, 1998.

Mourning Cloak Life Cycle*
November-February:
Adults hibernate.
March:
Adults feed,
find mates,
lay eggs,
and die.
April:
Larvae hatch,
eat, and molt.
May:
Pupae transform.
May-June:
Adults
emerge
and feed.
July-August:
Adults estivate
(rest in a safe spot).
September-October:
Adults feed.
*Timing is approximate.
It may vary from place
to place and year to year.